HOMEGROWN HOPS

An Illustrated How-To-Do-It
Manual

by

David R. Beach

HOMEGROWN HOPS-
An Illustrated How-To-Do-It Manual
by David R. Beach

Copyright © 1988 by David R. Beach

Library of Congress Catalog Card Number: 88-92165

ISBN 0-9621195-0-4

Printed in the United States of America

Published by David R. Beach
Reveille Farm
92984 River Road
Junction City, Oregon 97448
Tel: (503) 998-3369 or (503) 998-3643

Direct all inquiries/orders to the above address.

The views and conclusions expressed herein are those of the author. The author and publisher assumes no responsibility for the use or misuse of the information contained in this work.

Cover design and all photos by author.

Second printing - 1990

HOMEGROWN HOPS
An Illustrated How-To-Do-It Manual

by
David R. Beach

LIST OF CHAPTERS

INTRODUCTION

We came about this not quite by accident, my wife and I. In our years of traveling, particularly overseas, while I was serving in the U.S. Army, we had developed our appreciation for good beer. And we had gained a general awareness that people, ordinary folks like ourselves, made their own beer at home. But that awareness had not translated into any plans and certainly not to any actual efforts at brewing. Service life with its regular moves to unexpected locations does not favor such undertakings. However, after my retirement from active duty and our return to the state of Oregon, we found ourselves the owners of a very small farm endowed with fertile soil and a whole lot of fruits and berries already established. I guess what happened next was inevitable, but not foreseen by me.

One day, in early August a few years ago, my wife announced that she had a wonderful idea about what we could buy ourselves as an anniversary present, for our wedding anniversary upcoming later in the month. She said we ought to buy the items necessary to give a try at home brewing. There was a brewer's supply store in the major city not too far from our farm, so we paid a visit there and soon had selected a most unusual anniversary present for ourselves!

1

Well, one thing leads to another. Several years have gone by, and we have progressed from using cans of syrup and bulk extracts, through partial mashing of ingredients, to now being all-grain brewers. Along the way, the idea of putting some of our good soil to work growing hops for our brewing habit took hold. We have felt it was only natural to want to have as much of the beer making process under our own control as possible. As a consequence, we believe we are making much better beer these days. We can enjoy the glow of satisfaction that comes from knowing that we have grown the ingredient that gives any brew its unique character, whether that beer be home brewed or the product of a commercial brewery.

We are on a quest for taste. And aroma. Progress is being made, we believe, as we gain skill from experience and learn from talks with other home brewers and reading. But we acknowledge that there is more improvement yet to be made. There is what I have come to call the "*German nth degree*" of aroma and taste that still eludes us. Occasionally, a batch will have this special quality to a limited degree. But not regularly. I believe that hops, in terms of the kind of hops, the rate of hopping, and the timing of adding hops, is the key factor to this "*German nth degree*" of taste. So the quest goes on for us. That too, is part of the fun of being home brewers.

The pages and chapters that follow, and the photos as well, represent my effort to put down in permanent form what I have learned from our on-going experiment in homegrowing hops. I must add a note of caution in the form of a disclaimer, however, and that is that what follows is based upon my own observations and the conditions of our farm.

There is no assurance of success for other growers and other locations. But in that respect, I should pass along something I read this spring in a newspaper article about growing asparagus. That author told of "heeling in" some excess asparagus roots just a few inches down under a rough-tilled layer of earth and mulch. This was done as a temporary expedient, just to "save" the roots for a friend, as the regular method of planting called for asparagus root to be placed in about eight to ten inches deep. Well, to the surprise of the lady gardener, these roots soon were out-producing the ones she had planted according to conventional wisdom. Her message from that experience, and what I offer to would-be homegrowers of hops, is that different is not always wrong. Some of what I shall suggest below is "different" than what is recommended by other sources. And some readers of the manual will undoubtedly do their hops differently that I do ours. "Different is not always wrong."

I have enjoyed doing the writing and photos for this manual, just as I enjoy growing hops at home and making beer with them. Hopefully readers will enjoy the results of this creative effort, and the better homebrewed beer they produce as well. Welcome to the ranks of homegrowers of hops!

David R. Beach
Reveille Farm
Junction City, Oregon

3

CHAPTER 1

Choosing Materials and Laying Out the Hop Yard

Available space for growing will often dictate the materials to be used and indeed the shape of the hop yard. The key is to use your imagination. Those odd-shaped and ill-placed spaces around most any property, usually hard to keep looking neat and orderly, can be useful sites for a few hop vines. Examples are along driveways, next to existing buildings, or along fence or property lines. Do not yield to the temptation of expediency and plant hops with the idea of using any pole supporting electrical wires for the hops to grow on. The risks of electrical shock are too great. Save such spots for low-growing flowers or an herb patch. Do, however, consider areas even containing rocky or junky soil. What must be considered always is that hops do require sunlight and water, so select planting sites that receive at least half a day of full sunlight, and that are reachable by your hose or sprinkler system.

Since hops require very little square area of ground surface, but a lot of room for upward growth, they are suitable for use as a living fence or as a privacy barrier or screen. Besides this useful characteristic, hops of course pay the dividend of a crop that is necessary for homebrewing. The satisfaction of knowing that you are using hops you have grown yourself, possibly as a "bonus" from putting an unused piece of ground to use or gaining a visual barrier, is nice indeed.

Sometimes it is possible to make use of an existing structure to eliminate the need for poles to

Using existing structure, but twine too close together.

grow hops. The end of side of a garage or large shed may provide enough height that top wires may be supported from extensions fastened to the eaves. The question is, how much height is "enough?" In my opinion, a vertical distance of less than ten feet is asking for headaches from vines growing together in a jumble that makes cone growth and successful harvest doubtful. Closely packed vines are an invitation to moisture-retention and insect infestation. Mildew and bugs are bad enough to control at best; better not to plant in such a way as to encourage their presence! After all, the goal is to be able to harvest hops that you want to use in homebrewing, not buggy, mildewy cones that go into the trash instead of the boiling pot and fermenter.

If no existing structure of adequate height is available, or in areas that are open, such as between a lawn and garden, or in a garden itself, poles are the convenient way to support a framework for proper hop growth and cultivation. My choice was to order a quantity of treated and shaped poles from a pole yard discovered in the Yellow Pages. I selected Douglas fir, shaped to be three inches in diameter at the small end, and eighteen foot long. The large end ranged from five to six inches. These poles were immersion-treated to resist rotting. I placed the poles in holes dug three foot deep, then tamped gravel around the base while checking for plumbness with a carpenter's level. I backfilled with gravel almost to ground surface, then finished off with soil. Tamp in firmly and be gentle until guy wires have been installed to aid stability. I pre-drilled holes in the top of the poles to take the hardware for top wire attachment. If the attachment hardware is put in place before the pole is erected, you can observe its orientation, to assure that the pole is correctly rotated before being tamped down.

End pole with guy wire at right.

Commercial sources no doubt offer various galvanized hooks and eyes that may be used for top wire attachment. However, I had on hand some quarter-inch chromed steel rod, so decided to make my own attachment hardware with a quarter-inch threading die and some bending in the vise with big pliers. The resulting product worked fine. Just be sure to cut enough threads to be able to take up the slack in the top wires after attachment. The wire I used for the top wire was Number 9 galvanized. Again, a bit of muscle with some pliers produced the needed eye to couple the wire to the home-made hooks.

Once the top wires are in place, the next task is to attach guy wires to the top of end poles and secure the guy wires to the ground for stability. My technique was to angle the guy wires at roughly 45°, being mindful of the hazard they cre-

Attachment hardware.

ate to people walking by the hop yard, and attach them with a turnbuckle to a section of three-quarter-inch water pipe I had driven into the ground. The pipe used for this "deadman" was a section about four foot long, notched and drilled at the upper end to take a bolt to secure the lower eye of the turnbuckle. I simply drove these deadmen into the ground with a sledgehammer, taking care to angle <u>parallel</u> to the angle that the guy wire would form. Installed this way, I have noticed no movement in these anchors despite heavy rains and winds up to forty miles-per-hour in gusts. The screw anchor devices sold for securing mobile homes would, I imagine, work as well or possibly better than my home-made solution. Or perhaps a corner of a chimney base or other solid structure will be conveniently available to you. If so, use it, but pay attention to the hazard to people that the guy wire will pose and safety mark it accordingly.

After wires are all in place and turnbuckles are drawn up to tension the structure, the twines for the hop vines to climb may be tied to the top wire. That wire will be around fifteen feet in the air. That doesn't sound like much until you climb an orchard or other stable ladder and realize that there is nothing but air to lean on! Position the ladder carefully so you won't have to reach so far as to lose your balance; this is no time for a circus-like high-wire act! I have used the synthetic binder twine off of bales of hay or straw for hop twine and find it excellent. It does not rot in rain or sunshine, and can last for many seasons. I position these sections of twine about two feet apart (three feet would be better) and twist a short piece of fourteen gauge bare copper wire around the galvanized wire either side of the knot, to keep the twine from shifting along the top wire due to wind and weight of growing hops. Be sure to leave each hop twine long enough to reach the ground with some to spare for tying to the anchor wire.

-- Laying Out the Hop Yard

So much for choosing materials, now on to the challenging question of laying out your hop yard. Many available writings treat this topic on a grand scale -- by the acre, in fact. Surely that is helpful for the beginner commercial hop grower, but home growers of hops don't need to know the number of hills per acre and how to rig trellises for such sized plots. The home grower has neither the inclination nor the need for that kind of precision. With this in mind, the suggestions that follow are of a more modest scale and reflect the idea that the home grower will make use of whatever space is available. This will often mean small spaces, or odd-shaped plots.

If you do have the luxury of open ground, here is the spacing I have found to be satisfactory. Twenty-five feet between poles, and no less than two feet between hills of like-variety hops. I have a hop row fifty feet long -- three poles worth. I have used one half for hops of the same variety and have those hills on two-foot centers. Over several seasons the hills have tended to grow together. Hop rhizomes do spread with time. But the unwanted growth can be controlled by pruning during the growing season, and by more vigorous cutting at the fall trimming of the old hop vines. In the remaining half of this row, I have planted two different varieties of hops. I kept the two foot center-to-center spacing of hills of the same

Better spacing of twine.

11

variety, but left a four foot space between varieties. I arranged the hop twines accordingly. At that, there is a tendency for growth to intermingle at the top wire. This mixing presents a real problem upon harvesting, as the varieties need to be kept separate, in the mind of most growers, for storage and use. And hop cones of different varieties do not grow with little labels on them -- they all look much the same. So, if more space can be allowed between different varieties, both as to hills and as to the twines, the identification task upon harvesting will be made much easier.

Much of what has just been said about spacing in an open area applies to those more confined, odd-shaped pieces of ground that may be potential hop yards. I have learned from experience that such places are an invitation to plant hills much too close together, resulting in real problems with variety mixing at harvest time. The desire to use every smidgen of space is natural and understandable, but fight the temptation! You will be avoiding real difficulty later on. I suggest that two foot centers for hills may still be used in these smaller spaces, but use that spacing only along one axis. Experience says that at least four foot spacing should be allowed on the other axis, to have room to maneuver a ladder for harvesting the hop cones. It is no good to grow the hops, then be unable to get a ladder close enough to pick the cones. Leave some elbow room to harvest the bounty of your efforts -- and to keep the varieties separated so that you know what you are using in a given batch of homebrew.

CHAPTER 2
How Many Hills to Plant?

It is certainly understandable that homebrewers, particularly ones with any degree of "green thumb" at all, or who grow even a small patch of flowers, will at some point decide that growing their own hops makes good sense. Indeed it does, both from the perspective of economics, and from the inner satisfaction of gaining control personally over one of the key factors in successful home brewing.

Once the decision is made to give hop growing a try, the next question logically is how many hop hills are needed to supply a harvest of adequate size. Perhaps a word of caution is in order at this point, based on personal experience. If one has become accustomed to using a given level of hopping with commercial hops to produce a satisfying brew, that same level of hopping will most likely result in a brew much too heavily hopped once homegrown hops are substituted for the commercial ones. This is, I suspect, because the homegrown hops are truly fresher than any hops available through commercial channels. And fresher hops seem to provide more effective bittering and flavoring, ounce for ounce, than less fresh hops. In my experience, it is wise to cut the amount of hops used just about in half for the initial brew with homegrown hops, then make adjustment in amount to suit one's taste.

But having sounded that note of caution, the question remains as to how many hills one should plant. Here is a rule of thumb that gives a start toward an answer: plant half as many hills of bittering hops as you plant of aromatic hops. Now, if a homebrewer makes a five-gallon batch roughly every two weeks throughout the year, and if a hopping rate of two and one-half ounces per batch is estimated, then approximately four pounds of dried hops will be needed for a year's brewing. Production per hill in ounces of dried hops is subject to many variables, not all of which are wholly under the control of the grower. The most obvious of these variables is the age of the hill, along with temperature of the growing season. Drought can be compensated for by irrigation, but too much rainfall can occur as well. And bugs and diseases complicate the picture. My experience leads me to believe that, on average, and once full production is reached for the hills (usually by the second growing season), four hills of aromatic and two hills of bittering hops will produce sufficient hops upon drying to satisfy the requirements for the brewing rate assumed above, about twenty to twenty-five batches per year.

If a grower prefers to do homebrewing strictly with aromatic hops, this suggests a greater number of hills will be required. In general, aromatic hops tend to provide about half as much bittering per ounce as hops used principally for bittering. This is not a hard-and-fast rule. As hop variety breeders manipulate the qualities of the new varieties they produce, the dividing line between what is a bittering and what is an aromatic hop becomes less clear. And what any individual homebrewer perceives as "right" for his or her own brews in terms of aroma

14

and bitterness is subjective indeed. Some of the newer hop varieties are said to be both for bittering and aroma. And some are very strong in the bittering alpha acids present. "Chinook" is an example of these new and very strong breeds that is able to perform both functions according to its producers.

Some of the older varieties of aromatic hops tend to be light producers -- "Saaz" comes to mind -- while others, such as "Hallertau" can be quite prolific. Consideration of production characteristics will affect how many hills need to be planted. A good source of information for your locality is to talk with other homegrowers and learn from their experience as a starting point for your hop yard planning.

One final comment, in the nature of a gentle reminder, about how many hills to plant. Crowded hop vines tend not to produce abundant hop cones.

Crowded hills.

Crowded hop vines are more susceptible to disease and bug infestation. So, try to resist the temptation of thinking more growth of vines is better for a big harvest. The contrary may in fact be true. Adequate space for air, light, and moisture -- both between hop twines and on each twine -- will pay good dividends in hops harvested that are usable, with a minimum of effort and worry.

CHAPTER 3
Hop Nutrition

All living things need an adequate supply of proper foods, and hop plants are no exception. But how much of what food? That is the great unknown for the home grower of hops. Commercial hop growers of course answer this question by use of scientific investigation. Soil samples are analyzed to learn what nutrients are available and in what quantities, then shortages are corrected by addition of needed chemical supplements. This is fertilizer, carefully blended and tailored to the needs of the particular hop yard analyzed. The same avenue is open to the home grower of hops willing to undertake the required investigation and expense. But many seeking to establish a modest hop yard to sustain their home brewing efforts will have neither time nor inclination to do so. What shall they do to assure proper nutrition for their hop plants? The answer that I have adopted is two-fold: first, adequate preparation prior to planting; and second, experience that comes from the observation as the hop plants grow. While this two-fold approach cannot provide the advance information of a scientific laboratory, it surely is within the most modest hop growing budget's means and can give the grower a greater sense of participation in the undertaking. Soil preparation prior to planting is a way to hedge one's bets by providing an initial supply on nutrients for hops to be planted the following spring. And observation during the growing season will provide after-the-fact information on what is lacking in the food supply. Correction can by undertaken then, and also used to guide dormant season improvements to the soil.

Now to get down to specifics. Hops need the "Big Three" -- nitrogen, phosphorous, and potassium -- plus trace minerals for healthy growth. As far as what I have called the "Big Three," there is no need for any mystery. These are the three basic nutrients for plants that most of us have bought, or at least seen advertised on fertilizer sacks at garden supply stores or supermarkets. The labels usually show a series of numbers, such as 10-20-10, or 22-0-0, in bold figures. Those numbers are the manufacturer's way of indicating the amount of the "Big Three" contained in the particular formulation of a given bag of commercial fertilizer. A glance at the fine print on the sack will confirm which amount is for what nutrient, and will also reveal what amount of trace mineral is present. For some, however, the use of a commercial product will not always be the proper solution. Whether one is a duty-bound organic gardener, or merely a person trying to use up what would otherwise be waste, organic sources of the "Big Three" may be an attractive alternative. This is manure. The precise amount of nutrients is not shown on any label, but the food value is present. So, if one is inclined to put in the additional work involved, certainly organic fertilizer can be used to give hop plants what they need to grow. Elsewhere I have mentioned my belief in the use of manure-straw as a mulch and food source for hops. Results obtained satisfy me, and I use other sources of nutrients only when necessary as shown by the condition of the hop vines as they grow.

Recently my wife and I paid a visit to a representative of a major agricultural chemical firm[1] as part of the research effort of this book. We discussed the topic of hop plant feeding at some length. This gentleman volunteered that indeed a mixture of organic fertilizer, such as chicken manure and straw, supplied the necessary nutrients for good hop growth except one key mineral. That one key mineral just happens to be frequently insufficient in the Willamette Valley, apparently due to being leached out by heavy seasonal rainfall. The mineral missing from the organic mixture that I favor is boron. It is one easily supplied from borate granules, but must be used with great caution, as too much is toxic to the hop plants. "A pinch to the hill," was this expert's advice.

This same source of good information provided data on the percentages of nutrients in commercial fertilizers blended for use on Haas Company hop yards in the vicinity. I will pass along those percentages, but with the warning that the significance of this data for the home grower of hops lies not in the quantities, which are designed for hops by the acre, but rather in the relationship between amounts of nutrients. The nutrients applied, in pounds per acre, are: nitrogen -- 90, phosphorous -- 200, potassium -- 190, sulfur -- 50, boron -- 3. This mix was "correct" for the hop yards where it was applied. It was based upon careful laboratory testing of the soil, and in established hop yards. Of interest

---

[1] Interview at Independence, Oregon, with Mr. George Boyce, a representative of J. R. Simplot Company of Pocatello, Idaho.

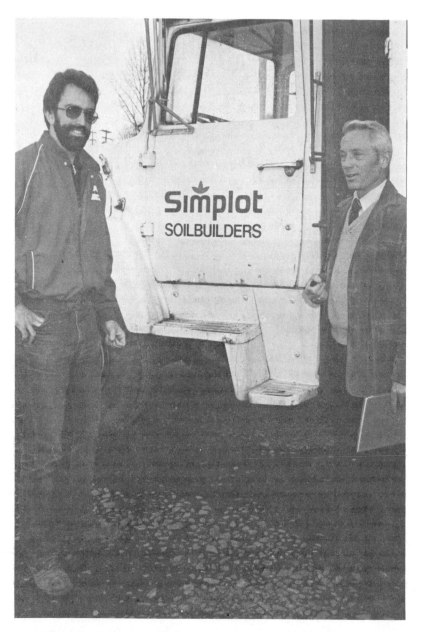

Mr. George Boyce, on left, with author at Independence, Ore.

is the relationship between nitrogen, phosphorus, and potassium. In contrast with home gardener sacks of commercial fertilizer, this tailored mix for these hop yards had roughly twice the amount each of phosphorous and potassium as of nitrogen. This is opposite to lots of home gardening mixes I have seen. And it may be a clue to what the home grower of hops can expect to have to use if commercial fertilizers are employed.

For those who prefer to avoid commercial fertilizers, there are other ways to supply essential nutrients, or unlock and make available those present in the soil, by application of materials that are natural and "organic" in the sense of meaning not from a chemist's retort. These include sparingly-applied wood ash, for potassium, rock phosphate, for low concentration but long-lasting phosphorous, and lime, to "sweeten" the soil's pH and thereby make available minerals that may be trapped, or locked up, in the soil when the pH is too acidic and thus not available for use by hop plants. These trace minerals include molybdenum, iron, calcium, and manganese. A visit or call to the local agricultural extension service, or a university's school of agriculture will provide invaluable advice on local soil conditions and how to correct problems. This source of good counsel should not be overlooked by the home grower of hops. Usually there is no charge for information provided and some publications may be free as well.

Underlying this discussion of hop nutrition is a point of importance that may be overlooked by some readers. For that reason, I will state clearly that two goals need to be kept clearly in mind by the home grower. These goals are quite compatible, but distinct: first, one needs to correct the soil of the hop yard for what is missing; and second, one needs to replace that nutrients that are used by the hop plants as they grow. Put another way, a person must supply what is not present at the start of the hop yard, and in addition, as a regular task during the entire life of the hop yard, the grower must replace what the hop plants take out of the soil each growing season. Neither goal may be neglected without paying a price in reduced harvest. And soil that is "made right" by adding materials for the first year may be in shortage again in future years simply because the hops have exhausted some element. All of which brings me back to a point made on several occasions, namely, to keep a good eye on your hops as they are growing for clues of problems that are beginning. Early attention will avoid major crises later and possibly the ultimate disappointment of a nonexistent harvest.

CHAPTER 4
Hop Hazards

For lack of a better idea on how to organize the following material, I choose to do so under the headings of natural hazards and man-made hazards. In truth neither category is exclusive, as one may do certain things to reduce or eliminate natural hazards, and those directly attributable to man are often remedied by forgiving Mother Nature.

One must use the soil and space that is available for growing hops. The proper application of imagination can result in better use of space available that at first glance seemed possible. Some ideas in this connection appear elsewhere and will not be repeated now. What is important, for present purposes, is to emphasize that even though space for growing hops seems unavailable, with good planning and imagination, a determined person will come up with a spot or two worth trying even though not ideal. Once those locations have been identified, attention needs to shift to the quality of the soil at the places of interest.

. While it is true that hops are generally a very hardy plant -- perhaps a reflection of the character of a relative of hops, the obnoxious wild nettle -- poor soil quality will result in reduced growth and harvest, and may place the hop plant at greater risk for disease and pest infestation. Something can be done to improve poor soil, and the wise grower, or more correctly, soon-to-be grower of hops, should consider investing some time during the fall and winter to this task. Test kits are sold in most gardening supply stores to allow one to test soil for pH, the balance between acidity and alkalinity. The standard scale goes from one to fourteen, with seven being neutral, and values less than seven indicating acidity. Hops are said to do best in a pH range from 5.5

Use available space.

to 8.0. I suspect that being on the lower side of this range is better than being towards alkalinity. Garden supply stores can offer ideas to correct severe imbalances of pH. Agricultural grade lime and wood ash both will "sweeten" soil that is too acidic, but must be used sparingly. They also provide for release of valuable soil nutrients needed by hop plants.

Certainly someone approaching the task of soil

preparation very seriously can go to the expense of having the soil fully analyzed by a testing service. In some cases that might be a wise investment. But if you know that your soil is capable of growing modest vegetables or flowers, chances are that major adjustments will not be required. Soil preparation for future hop hills may involve no more that working in well-rotted manure in the fall, with the addition of a dusting of lime, and letting the spot winter over to meld these materials into the soil. I have had good results with such efforts. A word of caution is in order, however. Manure or compost that has not fully digested ("rotted") requires nitrogen for this process to complete. Thus, if one adds such unready materials to soil, the result may be to deplete the available supply of nitrogen and produce poorer soil. That of course is opposite to the goal of preparation of the site to grow good hops. Careless preparation is probably worse than no preparation at all. The objective is to put some time and effort into being ready to plant hops in the spring, so that the likelihood of vigorous, healthy growth is increased.

Another natural hazard that deserves some thought over time is wind. Most regions will have a pattern of prevailing wind during spring when young hop vines are the tenderest and vulnerable to being snapped off. Even later in the growing season, as the vines reach full height and start sending out laterals that will bear cones, wind can tear leaves and break the cone-bearing members. So, to the extent possible, try to locate sites for planting to provide a wind break for your hop vines. This can be in the form of a building, a fence, or other vegetation such as shrubs or trees. Yield will be much better from vines screened from heavy wind. If this simply is not possible for your location, just realize that some loss in production is inevitable for the vine bearing the brunt of the prevailing wind.

The next natural hazard worthy of some advance thinking is water. Hops tend to be both thirsty and hungry, but both water and food can be overdone. Certainly thought must be given to how one will provide adequate water when natural rainfall is insufficient. The opposite side of that coin, and the point of concern here, is that thought must also be given to adequate drainage for hop hills. Hops do not need to stand in mud. Too much water standing at the hill encourages disease, rot, and eventual death of the hop plant.

If the locations chosen for hop hills are at risk for inadequate drainage, one may try modification of the soil by addition of sand, gravel, and humus to foster better natural drainage, or, one may choose to build up a small mound of soil and grow the hop hill on a slight elevation. Sometimes both methods will be needed to allow use of a given space.

I must now take a page from my "lessons learned the hard way" mental file and warn of a less-obvious natural hazard to good hops. This is shade. In the older of the two hop yards at our farm, I had noticed a gradual decline in production of hop cones from certain hills. It was to the point that I was having doubts about harvesting an adequate supply of one of the aromatic hops, and that was grounds for concern. I tried more fertilizer, checked for bugs or disease, and said unkind things about our prevailing north winds in the spring. But these efforts failed to solve the apparent shortfall in production. Then I noticed something, obvious but earlier neglected by me. The willow tree adjacent to this end of the hop yard had been busily growing, year by year, unmindful of how its increasing bulk was masking more and more of the hop yard from adequate sunshine. So, out came the pruning saws, lopper, and chain saw. And, within a month of this sudden surgery, sunlight had restored the hop hills in doubt to full vigor and my worries over an inadequate harvest had evaporated. So, lesson learned: watch out for unwanted shade in the hop yard. My rule for now is to stand on the hill (during dormant season, of course) and I must see sky unobscured by branches straight up. Otherwise, out come the tree surgery tools and space is made!

Now, some words about natural hazards one need not worry about (at least, in my experience to date). The good news is that birds and slugs do not seem to like hop plants. Other critters sure do, and more on that elsewhere. But at least it appears that the home hopgrower need not be concerned that slugs will eat the new shoots as they emerge from the cold spring soil, or that birds will do more than occasionally annoint a few hop leaves. Note that I have said nothing about rabbits or deer. I do not know whether either likes hops or not. I suspect they may, so if your growing area is frequented by such natural species, a bit of advance protection may be in order.

The man-made hazards that come most easily to mind are these: kids playing in the hop yard unmindful of what is under foot; ditto for dogs, cats, or if one's fences are inadequate, domestic livestock. Of course, the home hopgrower himself or herself can be the hazard, by walking on unmarked hop hills just as the new growth is trying to emerge, or by being too vigorous in handling new growth being trained onto climbing twines, or by the worst man-made hazard of all, neglect. Both too much and too little care should be avoided: too much water or food; too little watching to see how growth is progressing. Hops may remain alive and even produce without regular watering and attention to bugs and disease, but the cones for harvest will be fewer, smaller, and less appealing for use in the brew pot or fermenter. A balance needs be struck between the extremes of unfounded worries over every little change in growth, and yawning neglect. Remember, good growth and proper nutrition are the best preventive against harvest disappointment. Keep an eye open, regularly, to know what is happening in your hop yard. Treat problems early so that you may be happy in anticipation of a good harvest leading to a better home brew!

CHAPTER 5
Propagation Methods and Timing

Traditionally, hops are propagated by means of rhizomes. Experience has shown me, however, that this is not the only means to grow hops. But more on that later. First a few words are needed about the rhizome. What is it that this funny word describes? How do you know a rhizome when you see one? And most importantly, how should a hop rhizome be planted?

Rhizome is the name given to a specialized portion of the plant stem that grows underground. It is distinguished from roots in that it stores a food supply for new growth and has buds or nodes on it that can grow to become the above-ground portion of the hop plant.

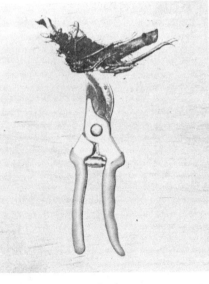

Hop rhizome.

29

The actual roots of the hop plant grow out of the rhizome downward and laterally, and are quite slim in comparison to the rhizome itself. The rhizome is usually rather gnarled or lumpy and in color ranges from tan to medium brown. With its fringe of roots and a few pale buds protruding, the rhizome reminds me in appearance of a slightly muddy brown shrimp. In length a rhizome may range from about three inches to nearly a foot, depending on how the supplier has trimmed it. The portion exposed by trimming should be very close to white, sometimes with a light brown center core visible. The rhizome should be firm, like a good potato, not withered or dried out.

The literature I read before planting my first hop hill was not very helpful on the most basic question of how a rhizome should be planted. In fact, the one brief mention I can recall said "vertically" and by my experience that is not true. What works for me, and is in harmony with what I have observed in hop hills that I have dug in to expose rhizomes after the hill is well established, is that rhizomes should be planted horizontally, parallel with the ground surface. I dig a small trench in prepared soil about four inches deep, aligned with the axis of the top wire between the hop poles, and place the rhizome in the trench with buds pointed as close to up as possible and the roots spread out to either side. Two rhizomes may be placed in one hill if desired. Then cover the rhizomes with earth and firm the soil by hand. I usually add a cover of manure straw, which serves to identify the location of the planted rhizome, keep the area moist, and provide a source of food once growth begins. Do not be concerned that the mulch of manure straw will prevent the new shoots from emerging; they are quite capable of working their way through mulch so long as it is not allowed to dry out and form a hard crust. In fact, the mulch provides a good way to satisfy natural curiosity whether the new hill is growing or not. Just keep an eye on the mulch over time. The new shoots will gradually push the mulch up into a small peak just before breaking out into the open air. So, "pimples" in the surface of the mulch merely means that growth has started and soon, new tender hop shoots will be out in plain view.

So much for the traditional way to establish new hop hills by means of rhizomes, what about other methods? My experience has shown that rhizomes are not the only way to propagate hops. I have used, successfully, two other methods and can detect no difference among the methods in the hop cones produced by the plants. Now it is true that hops can produce seeds, but I am not suggesting that planting seeds is a satisfactory way to propagate hops. Indeed, most writers I have consulted on this subject say that having both the male and female hop plants in a hopyard -- the only way to end up with fertilized seeds in the hop cones -- is not a good idea at all. The reasons given are that most brewers do not want to use hop cones containing matured seeds in their brews, and that the presence of seeds cuts down marketability of the hops as a consequence. I suspect another reason for not wanting male and female hop plants present is the undesirable risk of cross-breeding between varieties, something best left to the care of those who seek to create new strains of hops for the rest of us. The result of all this is that most hop yards consist of female plants only, and propagation must be done by other than seed. I have found two ways of doing so, beyond the traditional use of the rhizome.

Hops may be grown from the lower sections of the

above-ground vine, saved from the previous fall/winter trimming of the old vines. One should select a section that displays buds ringing the vine at

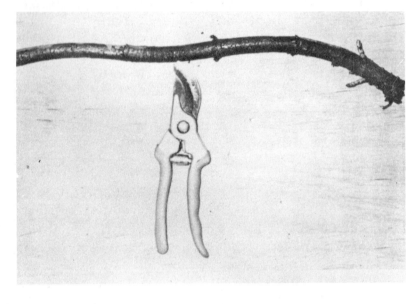

Vine section with buds.

the points where the leaves attached. Such rings of buds appear about every six to eight inches along the lower two to three feet of vine, especially on the sturdier vines about the diameter of a person's little finger or larger. From sections preserved over the winter, cut pieces long enough to have two rings of buds on the portion, and place in a trench in the same manner as done with a rhizome. Growth may not commence quite as quickly as with a rhizome, and indeed, some portions may not grow at all. It has been my habit to put two sections cut to length as just described in a trench, side-by-side, for each new hill. This method has succeeded very well for me. By the second growing season, I cannot tell from production of cones which hill was started by rhizome and which was started by vine section. Just

Vine section in trench.

be sure not to allow the vine sections to dry out over winter if you decide to try this technique of propagation. The cut portions used to start new hop hills must be alive, showing near-white inner fiber at the trimmed end, and a bit of green at the buds. Vine sections may be preserved over the winter in a compost heap or leaf pile, or by heeling in, in a shallow, marked trench.

Another less-reliable but still useful method of propagation I have used is to make use of excess new shoots from an established hop hill. When these older hills are just beginning renewed growth in the spring, many more new shoots will appear than the hop grower will need or want to train up the hop strings. Keeping unwanted growth trimmed at ground level will become a task to be performed at weekly intervals during the time of most vigorous growth. I have taken particularly strong specimens, and instead of trashing them, used them to start a new hill or augment an existing hill of the same variety with new growth. The technique I have used is very simple. I have an old skewer on hand from a long-discarded barbecuer. It is a square rod, about

Propagation by shoots.

pencil diameter and three feet long or so. I poke a hole with this skewer about a foot deep, then insert a just-trimmed new shoot, making certain that I have placed at least two leaf nodes below ground level.

Since the hole made is quite small, a bit of pressure with a thumb or finger closes off the moist soil surrounding the shoot. Shoots so planted will start to droop within a day, reflecting their quest to get needed moisture I suspect. So keeping the new hill well-watered

Placing the shoot.

is important if spring rainfall is insufficient. In about a week, you should be able to tell if the transplanted shoot is taking off with growth of roots or has failed to establish itself. I have had about a fifty percent success rate with this technique when done early in the growing season, to coincide with the most vigorous growth period. Later in the season such efforts at propagation have not worked well.

By using these alternative hop propagation methods, the home grower of hops can expand production at no out-of-pocket cost at all. There is the initial expense of purchasing rhizomes from a reliable commercial supplier, or obtaining them from another grower, to establish the variety of hops desired. But once that initial cost is invested, the home grower can increase the number of hills at will, by taking rhizomes from established hills (I recommend no sooner than the second full growing season), or by using the stem section or new shoot methods I have mentioned. All that is required is time and patience; soon the home grower will have more propagation stock than can be used, and will be seeking to pass this stock along to others.

-- <u>When to Plant</u>

Our little farm (and its hop yards) is located in the southern Willamette Valley in western Oregon. This is soggy country, cool and damp for about nine months each year. Heavy snows and hard freezes are nearly unknown in this region, and by February, the sap starts rising in gardeners just as it does in early season plants. All this is to denote a bit of caution for the reader: What I am about to say on the matter of timing for planting hops must be adjusted to the climate of the user of the information provided. The one bit of advice I shall offer that might lay claim to universal value is this: If you are unsure as to correct planting time, err on the side of being early rather than being late. And I have found that commercial suppliers of rhizomes frequently recommend a later time to plant than my experience says should be used. If the mulch technique mentioned earlier is adopted, a newly planted hop hill can withstand freezes of 20°F without damage, so why not be sure you are early enough?

For where our farm is located, I believe that February is the month to plant. March is still satisfactory, but April is just too late. Growth will have already started by then, and the strong pulse of that early growth will be lost to any hill started then. As a rough rule of thumb, hops can be planted as early as one would plant early peas -- and I have heard it said that peas should be in the ground when Lincoln's Birthday/Washington's Birthday (now called Presidents' Day by some) rolls around. So, if you are dependent upon a commercial supplier for your initial stock of rhizomes, get in contact early and insist on early shipment of your stock. Once your shipment arrives, keep the rhizomes refrigerated (NOT frozen) in a plastic bag to prevent drying, until planted.

Those long-practiced in gardening will no doubt say that there is a serious flaw to what I have just stated on timing for planting. They will note that at the time suggested, the soil is far too wet to be worked by tilling or spading. That indeed is true. However, one can prepare the intended spot during the previous fall by adding humus, lime, or other materials to ready the soil. Actually it is my experience that very little soil preparation is necessary. Hops are hardy and can prosper in soil where other plants will not. So long as competing weeds and useful plants are removed from the immediate area of the new hill, hops will demonstrate amazing stamina despite the lack of carefully prepared soil.

CHAPTER 6
Hop Diseases and Pests

Sooner or later, most every homegrower of hops will get the feeling that he or she is the helpless victim of unseen enemies that are out to thwart the grower's best efforts to produce a good harvest of hop cones. The feeling may be based upon observation of withering hop vines where just a few days ago lush growth existed, or it may arise vicariously from having done some immersion reading on hop diseases and insect problems. Whether based in fact, or merely in the mind, the remedy for this feeling is much the same. Information will reduce these unseen enemies to understood challenges and (hopefully) manageable frustrations, solvable by patient attention. The material that follows[2] is an effort to cut these unseen enemies down to size and make them recognizable to the homegrower. It is not an exhaustive study of the topic, as that treatment is better left to those specialists in plant pathology at major agricultural research facilities who spend their lives doing so. I only can offer simple tips and

---

[2] A portion of the information provided in this chapter was derived from an interview at Independence, Oregon, with Mr. George Boyce, a representative of the J.R. Simplot Company of Pocatello, Idaho. Permission to quote granted by Mr. Boyce.

tactics, without fancy latin names of bugs and diseases, leaving those curious for greater detail to pay a visit to their local agriculture extension agent or university.

The way I see it, a sound plan to protect against hop disease and pests involves three elements: prevention, natural treatment and control, and as a last resort, chemical means. I will try to give emphasis to the first two elements, but will mention tips on chemical means within my knowledge for those who will want to make use of "last ditch defenses." A word of caution at this point is clearly in order. Commercial growers use chemicals that are toxic in the extreme. They do so with protective gear and clothing (and scientific knowledge) not readily available, or affordable, to the homegrower. These chemicals are effective but are very dangerous to apply and store, and must be disposed of in accordance with E.P.A. and state rules. The chemicals I shall mention hopefully do not include these very dangerous ones used by commercial growers. Still, be cautious and respect warning labels for use and disposal.

The first line of defense is prevention. Healthy vines, well fed and watered. A hop yard clean of weeds and last year's growth is the starting point. Certainly hop vines or leaves known to have been diseased or bug-infested should not be allowed to remain in the hop yard, but should be disposed of properly. That is, not in the compost heap where the problem can come back to haunt in the future, but in trash bags to be taken off the property. Weeds allowed to grow around the hop hills can be harbor disease and bugs as well as robbing needed nutrients from the hop vine. So there are sound growth reasons for weeding beyond the satisfaction of a good-appearing hop yard. One should also avoid doing things to hops in circumstances likely to lead to problems. For example, fungus grows well in moist, dark conditions. Therefore, avoid watering the leafy portion of your hops so late in the day that water will remain on the vines after darkness. Certainly rain falls by night, but why make matters worse by watering so late that fungus growth is encouraged when it has not rained? I try to allow two hours from the end of watering until full darkness, so that vines and leaves can air dry. And, as noted elsewhere, I try to remove nearby competing vegetation that shades my hop vines. Open space for circulation of air and direct sunlight seems to discourage the growth of fungus and spread of bugs.

As mentioned above, information is the key ingredient to counter the enemies of your hops. The homegrower must know how to recognize these enemies before any control or treatment can be undertaken. The sections to follow on diseases, pests, and deficiencies will emphasize recognition as the vital first step in combatting these enemies.[3]

-- Diseases

Mildew, wilt, and virus infection are the three main culprits in the disease category. Both mildew and wilt can kill hop vines; virus infection tends to slow, stunt or distort growth and may eventually kill the hill infected.

The mildew fungus, usually descriptively called downy mildew, is possibly the worst challenge to the

----

[3] The J.R. Simplot Company wall poster titled, "Hops-Nutrient Deficiency, Disease and Insect Damage Symptoms," copyright June 1975, has been an invaluable source of information for this chapter and is gratefully acknowledged by the author. Permission for reference use, with attribution, was obtained.

homegrower of hops, due to the suddenness with which it can appear and the speed with which it spreads. Literally overnight healthy vines can be afflicted. Regular inspection is essential to avoid this problem. In springtime, just as new shoots begin to appear, mildew can attack the new growth. Some shoots will be normal, but others will appear spike-like, be brittle or stiff with leaves curled under, and have a coating of whitish-gray film. Such growth should be cut free, placed in a sack and disposed of outside the hop yard in the garbage can. The fungus that causes mildew is very hardy and can winter over in the soil. Be careful to pick up all infected pieces for disposal. For chemical control, try an early spring drench with Subdue, the gardener-size package of the product Ridomil, being careful to follow directions. This fungus can hit later in the growing season as well, especially if the weather is cool and moist for several days. On vines that have already started climbing, the infection will appear initially as a whitish, fuzzy film, often in irregular patches on the underside of leaves. Quickly it will grow to resemble the result if someone had sprayed the vine with that white Christmas "snow" aerosol spray. In a matter of days, the leaves will yellow, then show angular browning or blackening. I have used the fungicide captan against this growing season outbreak with some success. The product called zineb is also said to be effective, but bear in mind that zineb must not be used within fourteen days of cone harvest. Some growers advise the removal of the lower leaves of the vines, to a height of two feet, to slow or prevent the spread upwards of the mildew fungus. I have not tried this technique so I do not know whether it is effective.

Another fungus causes wilt. This disease is

formally labeled verticillium wilt and appears on hop plants as a yellowing and dying of leaves, from the base of the vine upwards. The afflicted leaves will develop a striped appearance, with bands of dull-green colored ("cooked spinach" looking) tissue alternating with yellow bands, parallel to the veins in the leaf. Such growth should be removed and put in the trash for disposal. Fungicides used against mildew may be effective for wilt as well; contact your county agriculture agent for advice on products.

Various virus infections can occur in hop plants. There is some question as to how much harm results from virus damage. Reduced harvest seems to be the most obvious consequence, although more serious risk to the perennial growth part (crown) of the hop plant cannot be ruled out. Signs of virus infection include yellow ring spots on the surface leaves, upward curling of leaves, distortion of leaf shape, gradual yellowing of whole leaves from mottling that spreads, and sometimes a failure of vines to climb. Afflicted parts should be removed and disposed of in the trash.

-- Pests

Lots of different bugs, worms and "critters" inhabit hop plants. A few are beneficial; most are not. The major offenders will be noted below, and one very good bug that serves as a defender will be given special mention.

Probably the most common bug pest on hops is the aphid, also known to some as aphis. This translucent pale green bug infests many garden plants and flowers, and will migrate from plant to plant in search of juices from tender growth. Ants "herd" aphids and use them as cows, for the secretions they produce. Aphids are easily seen and can become so plentiful as to nearly obscure the stems of plants. Cool weather favors aphid infestation. Left untreated, severe infestations will rob a hop vine of nutrient juices necessary for growth and result in defoliation and death of the plant. A natural enemy of aphids (and obviously a good bug from the homegrower's viewpoint) is the ladybird beetle, commonly called the "ladybug." They eat aphids and do not eat hop plants. But the question is how to get these familiar orange flying bugs to stay put on one's hop plants. One solution I have heard but not yet tried sounds quite reasonable and goes thusly: locate a swarm of ladybugs, say on the siding of your house or under a window sill, and capture them in a glass jar. Put the glass jar into the refrigerator for a day (NOT the freezer), to force the ladybugs to exhaust their body's food supply just to keep warm. Then, release the now thoroughly chilled and hungry ladybugs onto your hop vines. Instead of flying off elsewhere in search of a mate, they will, according to this theory anyway, immediately seek out food. The aphids on your hops will be depleted in short order, and hopefully the now-happy ladybugs will do their procreation in the hop yard, to leave their offspring handy for the next aphid attack. Natural defenses against hop pests, such as the ladybug, are wonderful indeed. But unfortunately, a ready supply of ladybugs for capture may not be available. In that event, chemical means may be the only answer to a large aphid problem. I have used diazanon as a spray with good results. The major concern with it is that fourteen days must lapse between spraying and harvest time. So, bear that limitation in mind if you plan to use chemical warfare against aphids.

The next pest for consideration is the spider mite and I am not aware of any natural defender against this pest. This critter prospers in hot dry weather, and is so small that detection is difficult. The evidence of spider mite presence, however, can be seen. This is in the form of webs on the underside of leaves, and light colored spots, like freckles, on the upper leaf surface. The freckles are the result of the injury caused when the spider mite pierces the leaf to suck out juices. The webs function as a protective cover for the tiny mites, and make it difficult to spray or dust to kill them. A sure but late sign of spider mites is the appearance of rusty-red cones, and in severe infestations, defoliation of the vine. Sulfur applied as a dust is said to help control spider mites.[4] But the most commonly mentioned chemical means that is very effective is Kelthane, a commercial-type weapon. It must not be applied within seven days of harvest, and hop plant residue may not be fed to dairy or meat livestock. Kelthane may not be as hazardous to use as some of the commercial systemic insecticides, but all warning labels should respected fully. It is classified as a miticide and is not considered to effective against insects, such as aphids.

Another rather common hop yard pest is the western spotted cucumber beetle. It moves from plant to plant throughout the garden, much as aphids do. It is about the same size and shape as the ladybug, but instead of being colored orange with black dots, it is yellow-green with black dots. These

---

[4] Sulfur, and other "old-fashioned" treatments for pests, such as oil and soap, are being given a serious second look by experts as offering less hazardous ways to control insects than modern chemicals. Aphids and mites both may well be susceptible to these older remedies. For further information, see the August 1988 issue of Sunset magazine at page 126, "Safer strategies for battling summer garden pests."

beetles feed on the tips of the hop vine and on cones. If they are present in such numbers as to become a problem, they may be attacked with diazanon, just like aphids. But recall, fourteen days must pass between last spraying and harvest.

Other bugs in hop plants include earwigs, various worms, and grubs in the roots of the hop plant. These pests, while not pleasant to encounter, are not usually a major concern as they seldom threaten the whole harvest. A broad-spectrum insecticide, such as diazanon, often will control them as well as more worrisome critters like aphids. And, cones chewed upon always can be discarded if only a few are affected.

-- Deficiencies

It would certainly be nice, and very convenient, if each deficiency, and each other malady affecting hops, had its own unique recognition symptoms. But that is not the case. Frustrating as it may be, symptoms of deficiencies often resemble those of disease or pest infestation. The slight differences that do exist allow the careful grower to track down, with some degree of confidence, what is threatening the harvest. By the process of elimination and practice, confidence will grow (and hopefully, the hops as well).

I will now go through a listing of various nutrient elements and list the key symptoms of deficiency for each.

Nitrogen: Dwarfing of new growth, light green color that yellows later in the growing season, leaf stems may show light red color.

Phosphorus: Leaves are smaller than usual and are a dull olive green that can fade to dull orange, some downward curling of leaves and leaf stems may become dark red, brown spots appear on the underside of leaves between the leaf veins.

Potassium: Downward curling of leaves with bronzing of tissue between the leaf veins on mature leaves, scorched appearance to younger leaves after becoming pale green, followed by ashen color and leaf fall.

Magnesium: Bright yellow color on tissue between leaf veins of lower leaves, progressing to dead leaves and early defoliation, symptoms spreads upward on vine, mottling of leaves as they yellow, then turn brown before dying.

Calcium: Leaves are smaller than normal and tip growth is weak, edges of the leaves show yellowing before dying.

Iron: Symptoms appear first at the top of vine with youngest growth, leaves yellow and veins of leaves are pale green, leaf tips may remain green while base shows yellow.

Zinc: Symptoms appear right after growth begins from ground surface, leaves roll upward and have glossy surface, followed by bronzing and browning, leaves are more narrow than usual, some yellowing of leaves and stunting of growth, rings of leaves are progressively closer to each other on vine until a cluster of leaves appears at end of vine.

Boron: Also shows early in growing season, with uneven growth and shoots that are thicker than usual, some shoots die back, leaves are small and distorted or wrinkled.

Molybdenum: White speckling on leaves and curling upward, as symptoms progress leaves show some yellowing between veins as speckles blend together.

-- Other Problems and False Clues

If similarities in symptoms for deficiencies, diseases and pests were not enough to lead to confusion, add in other, look-alike problems and the scope of the challenge to the home grower of hops becomes even more pronounced. Wind not only can tatter leaves, but because the edges and surface of hop leaves have tiny hairs and feel "rough", even rubbing together from wind action can produce surface injuries that have an appearance of more serious problems. Injury due to an unexpected cold evening and resulting frost can resemble other worries, leaving young hop growth stunted and leaves curled.

Not much can be done about these tricks by Mother Nature. One can but hope that the harvest of hop cones is not seriously affected.

Probably the best rule of thumb I can offer is to avoid over-reaction to first symptoms of maladies. Do take time to think through and eliminate unlikely causes or those things over which the grower cannot exercise control. Then start a corrective program proportionate to the size of the problem detected. And if nothing seems to help, simply accept the likelihood of some losses at harvest and look forward to a better next year.

Healthy early growth.

CHAPTER 7
Growing Season Maintenance

There are some differences between the care to be given to a new hop yard and what is done for an established one. But before discussing those differences, let us consider those activities that are common to both types of yards. These are the tasks of weeding and cultivation; watering, feeding, and watching; and early trimming and training.

In my experience, the home grower of hops is better advised to emphasize weeding than to attempt significant cultivation. What I mean by cultivation is digging, plowing, or rototilling. Such work has its proper role in hop growing, but it is at the time of preparation of the soil, not after the hop plants have been put in the ground. A hop yard, after all, is intended to produce year after year. The hills will gradually expand over several seasons, and indeed may have to be restricted by cultivation nearby lest they spread into areas where hops are not wanted. But at the hills themselves, cultivation must be limited or damage will be done to the plant crown that is the source of annual growth. It is my contention that so long as mulch is maintained around the hill, roughly in a circle of three to four foot diameter, to keep the soil moist and softened, there is no need for cultivation as such. The soil will be stirred a bit by weed digging, and that task will fully occupy the grower's time and attention. Remember, weeds can harbor disease and bugs, so they should be removed regularly. Weeding is also a good way to keep an eye on the progress of one's hop plants, for signs of bugs, disease, or nutrition problems.

Watering has been mentioned at various points in this manual. Right now, the emphasis is on regularity and sufficiency. Hops are thirsty plants. A lot of water is needed just to carry plant juices all the way up those circling vines reaching for the sky. And the cones that are the object of all the effort invested by the home grower are produced on the upper part of the vine, often starting at about five feet above ground level, then going upwards to the very tip of the vine. So do not neglect water as a key element to the production of a good harvest. In the spring when growth is most vigorous, natural rainfall may supply the bulk of needed water to the hop plants. For those growing hops in climates not endowed with ample spring rains, water must be supplied. I suspect that a drip water system may be the best for economy and deep watering. For those not inclined to invest in the hardware needed for drip watering, sprinklers of some type may be the best answer. I use the rainbird type of articulated sprinklers. And in the hottest, driest portions of the summer, I do water every day. But, as noted elsewhere, water should not be allowed to remain on the vines after sundown, due to the increased risk of mildew and other disease.

If a home grower of hops adopts the manure mulch technique that I have used with good results, feeding of the hops will occur each time the plants are watered. If plain straw (or even lawn clippings) is used for mulch, nutrition from another source will

Mid-April growth.

be needed. That can be manure added periodically during the growing season, or from chemical fertilizers. Just as hops are thirsty, so too they are big eaters. Some advise to supply a bit of extra nitrogen to the hop plants just as the burrs appear, signalling the start of cone growth. This clue for extra feeding must be noticed, however, and is a further reason for regular observation of the hop yard. If nutrients are to be supplied via chemicals, consider using a hose applicator rather than merely sprinkling the dry granules on the hill. Using a water hose applicator reduces the risk of "burning" the hop plant by too concentated doses of food and speeds the needed nutrients to the root zone beneath the hill where they may be absorbed. Remember

Hop burrs.

to keep track, however, of the garden hose supplying the water for application so as not to damage tender hop growth or other garden plants. If observation of leaf condition indicates the need for adding lime or wood ash, these too should be watered in after application to avoid avoid "burning" and aid absorption. The same is true for application of any trace minerals that observation reveals are needed.

Whether one is dealing with a newly planted hop yard, or one that has been in production for several seasons, there are common requirements for early trimming and training. The difference between required care for new as opposed to older hop yards is that of how much of what kind of care. In newly planted hills, any growth that appears is cause for great joy in the heart of the home grower. As a result, there is the tendency to let every shoot grow

Trim at ground level regularly.

and climb. That is surely understandable, particularly for non-vigorous varieties of hops, such as Saaz, which can be agonizingly slow in early years of growth. But one should fight this tendency to let every shoot grow. Select three or four of the strongest shoots and trim off at ground level the weaker ones. Force the strength of the rhizome to go into those shoots that you have selected to save. This will pay dividends later, at harvest time.

In established hop yards, the early growth is often astonishing to behold. If regular attention to trimming is not provided, a veritable jungle of growth will result. Harvest will be diminished and made much more difficult to do. So severe trimming, down to two or three shoots for "climbers" per rhizome is an essential chore that must be performed every two weeks or so. Let me say this again, for emphasis: Trimming must be done every couple of weeks. Those who choose not to do it will quickly understand why I have given emphasis. Early growth is surprisingly fast. Like a foot plus per week. In established hop yards, neglect at this stage equals a mess. Best to have your poles in place, wires strung, and twines attached. Because the twin to the chore of trimming is the task of training.

In the northern hemisphere, hop vines will wind themselves around any vertical object in a clockwise fashion, as viewed from above. This, I am told, is a result of the growing tip sensing the movement of the sun from east to west. Apparently the opposite is true for hops grown in the southern hemisphere. So, for any Aussies or Kiwis that happen to come across a copy of this manual, anticipate that your hops will grow "anti-clockwise," as the expression is in Mother England. The lesson to be drawn from this is simply to advise the homegrower that the emerging hop vines require a bit of training to know what to climb. Hopefully, this is the twine thoughtfully attached before growth begins in earnest. If twine and a bit of encouragement are not provided, the vines will seek whatever support can be found, including other hop vines. Again, the result of unattended growth is a mess. All that is needed is to gently wind the emerging vines selected to climb around the twine (in that clockwise fashion, of course) for maybe a foot or two. Once thus started, the vine will continue its spiral growth to the top wire and beyond. This may be some twelve to fifteen feet above ground level. It is my practice that once the vines selected for climbing have reached the top wire, I then relax a bit on further trimming of other shoots. There is a reason for doing this, as I shall explain shortly. But first, there is another important point to

Early training of vines.

mention under the topic of
training. This is how to
make an attachment for the
climbing twine at ground
level. I have tried several
ways, and have concluded
the simplest and most
satisfactory technique is to
use a short length of stiff
wire with a hook bent at one
end. I simply poke this
roughly foot-long upside
down "J" wire into the soil
at the place where I want a
vine to climb and tie the
loose end of the twine to it.
Leave a bit of slack, for

*Ground attachment
"J" wire.*

swaying in the wind. This provides a flexible base of
support for the growing vines, and is reusable from
one season to the next.

Now at this point I shall pause, briefly, to tackle a
myth. Or, what I believe to be a myth. This is the
advice offered by some that early growth from one's
hop hills should be trimmed off completely. This
suggestion usually is followed by a statement that one
should wait for later "deep" growth to emerge and
train that to climb. Frankly, my experience to date
does not agree with such advice. The only reason I
have heard advanced for cutting off all early growth
is the danger of freezing. That may be a good and
true reason, but I note the difference between frost
and freezing. Hops can withstand frost. And by
frost, I mean an overnight decline of temperature
below 32°F, with daytime temperature soon rising to
above that point. Now, if the weather is so severe that
it stays below 32° all day, or for several days in a row,

undoubtedly that would do some serious damage to newly emerged hop shoots. So, do a bit of reflection on your own local weather patterns, before accepting the myth of cutting off all early growth. I do not do so. I have found that this early growth is often the most vigorous, so why waste it? The goal, after all, is a harvest of good, flavorful hop cones for use in the brew pot, with a stock in the freezer for use until the next harvest. Go for it!

Which brings me back to a point noted earlier, and that is my habit of easing back on trimming once the initial vines have reached the top wire. I am unwilling to settle for a single picking of hops at harvest. I see no good reason for only picking once. Instead, I harvest the cones that have grown on the initial vines, usually beginning in June, then relax for about a month before picking the cones from later growth. That later growth comes from shoots not trimmed off once the initial vines have reached the top wire. These later vines are busy climbing to the top while the older ones are growing cones. So, once the new comers reach the top they too grow cones, and second, third and even fourth pickings are possible. But only if one opts for picking off of a ladder. Of course, if the older vines are cut down, secondary growth cannot wind its way skyward. For the home grower, ladder picking is feasible. The additional time involved is, in my view, more than justified by the opportunity for subsequent pickings and a more bountiful harvest. Commercial growers, using mechanical pickers, of course, cannot enjoy this opportunity for more than one picking. Just remember to be careful on the ladder; you will be a long way up from the ground to reach those cones at the very top wire!

CHAPTER 8
Harvest Time '

Harvest time is the justification for all the hard work, patience, care, and attention invested by the homegrower of hops. This is the time to gather the reward in the form of delicate green cones of fragrance so essential to a good beer. If Mother Nature has been kind, and if the grower has been diligent, the beginning of harvest time should be a period of great satisfaction and anticipation. Soon those cones will be available for the brew pot, with enough on hand to store in the freezer to meet brewing requirements until the following harvest next year. Good, clean, bug-free cones of pungent loveliness! That is the goal towards which all effort has been focused. Now that the moment of beginning is at hand, savor it, and devote the time required to do a good job and maximize your harvest.

When should one expect to begin to harvest hops? In the part of Oregon's Willamette Valley where our farm is located, harvest can start as early as late June on our Hallertauer hops. And, thanks to the techniques I have mentioned elsewhere, we pick hops through the month of October. Hop harvest for us lasts roughly four months, with some picking usually every two weeks or so. There is, of course, some minor variation to this schedule from year to

61

year, depending on weather and growing conditions. As to when I know that the time to start picking has arrived, I offer these suggestions from my experience: pick a cone at random, neither the largest nor the smallest, the lowest nor the highest. Take a sharp knife and split the cone on a board from tip to base and examine the two halves. If the cone is ready to pick, you will notice a bit of yellow dust or powder at the base of each scale of the cone. The yellow powder should be a fairly dark shade of yellow,

Checking for ripeness.

not unlike the yellow of highway lines, and it should have a pronounced hop ódor. If the powder is still a pale yellow, or is lacking in fragrance, better wait a week and try another testing.

The second method I use to determine readiness for picking is that of feel. There is a difference in the feel of a cone ready to pick as opposed to one still too green, and once the difference is learned, it is easy to know just by feel when to start picking. A yet-green cone feels slightly damp to the touch and has a softness to the scales; when squeezed slightly, it stays compressed upon release. By contrast, a cone ready to pick will feel papery to the touch, not damp. It's temperature to the hand will be no different from the air temperature, not cooler as a green cone usually feels. A ready cone will spring back upon light squeezing and seem light to the touch. As a final clue, a picker's hands will quickly take up the smell and slight stickiness of the yellow powder (a resin) if the cones being picked are truly ripe. This occurs within the first five minutes of picking. By using both of these techniques, a homegrower can determine when the harvest may begin. Bear in mind that it may occur that vines mature at different times, even side-by-side. This creates a problem for the commercial grower who uses mechanical pickers and removes the vines as part of the process. For the home grower, however, the later maturing vines may be held over for a week. This is one real advantage of hand picking. And, if a grower adopts the off-the-ladder method of picking, where the vines are not cut down, it really doesn't matter if a given vine is picked one day or another. Picking will extend over time in any event, so a leisurely pace is to be anticipated. Besides, your dryer may be full anyway!

The flip side of the when-to-start-picking coin, is the matter of when to stop picking cones. I have in mind those that you may not have had enough time to pick when they became mature, or for picking of which the weather did not cooperate. (By the way, my advice is to do the picking only on dry, calm days and certainly not during any rain or mist.) Mature hop cones will gradually start to show tan on the tips of the scales (actually the petals of the hop flower, which the cone truly is). This tan shade will darken to medium brown and slowly replace the green of the mature cone. I will pick a few cones just starting to show tan, but reject any that have gone very far into that stage. I do not pick and keep any that have turned brown. Frankly, those do not appeal to me as proper material for my brew pot. It is my belief that while under-ripe cones may not have the full flavor potential of ripe cones, they are far less objectionable

Harvest Time !!

for brewing than are the past-prime tan tinted cones. I suspect the older cones may be bug-infested, or otherwise off the mark in quality. Those I discard.

A word or two on picking technique is in order at this point. When possible, use both hands, one to hold the vine while the other detaches the cones. Sometimes the location of the cones and consideration of preserving balance high on a ladder will dictate use of one hand only to detach the cones. Simply use the thumbnail in that case, to separate the cones from the lateral stem of the vine by pinching firmly. The point is to avoid unnecessary injury to the vine. After all, immature hop cones may be dependent upon the vine to reach maturity.

Now I have made much of the off-ladder method for picking hops, with the reason being that "once is not enough," and that more than a single picking is indeed possible for the careful hop grower. However, I acknowledge that there may be perfectly valid reasons for certain growers not to adopt that method. In such a case, the grower will likely cut the vines off at ground level, remove the twines around which the vines have been growing at the upper wire, and do the actual picking of the cones away from the hop yard. In this circumstance, the grower will have to screen out unwanted cones before drying. The far too green cones, and those with too much tan showing, will be discarded as part of the price the grower is willing to accept to have the harvest over in one session. As with all handling of hop cones, gentleness is a good watchword to keep in mind when carrying the cut hop vines to the picking site. That yellow powder will sift out if the cones are handled too roughly, and its loss decreases the usefulness of the hops in the brew pot. So every effort should be made to keep that lupulin powder in the cone, where it will be available to flavor that brew!

If the grower opts for the ladder method, the question is what kind of ladder is needed. Certainly, a safe one is the obvious first consideration. Although "ten feet" doesn't sound like much distance at all when your feet are firmly planted on the ground, that same "small" distance becomes something much more significant when viewed as the space below your body as you sway on a ladder that far above ground. An extension ladder may be just the ticket for painting the upper reaches of your house, where it may be safely leaned on the exterior wall or the eaves. But your hop yard likely will not have the sturdy support necessary to use an extension ladder safely. The upper wire is not strong enough, for sure. While the poles (or other support means adopted to hop yard use) may be able to provide support for an extension ladder, the range of picking will be severely limited to the reach of your arms from one position. So, a good ladder for hop picking needs to be self-supporting. Step ladders are of this sort, but few people own a step ladder that is tall enough for use in the hop yard. The answer is to have an orchard ladder of about fourteen foot size available. These three-legged contraptions may seem the devil's own invention when it comes to moving them to, from, and within the garden area. They are awkward to carry and heavy, and can "wipe out" valuable plants and fixtures when due regard is not paid to what sort of an arc each end is moving in. But the orchard ladder is by far the handiest ladder from which to pick hops. Sure it flexes and wobbles, and one must be careful to keep the body within the bounds of the triangle of the legs to avoid falling over. But it allows the picker to get right in among those hop cones, to pick them off without stretching and reaching.   My picking technique is to use a clean fiber-mesh onion sack tied to a step on the orchard ladder about chest high when picking the highest cones.   Then as I harvest a hand full, I dump them into the sack with only a glance to assure that I have found the opening.   As I work down the ladder, I remove the sack and attach it to lower steps.   Avoid letting the sack with its precious cargo touch the soil or other contaminants.   Enjoy your harvest; you have surely earned it!

CHAPTER 9
Drying, Packaging, and Storage

Once the hop cones have been harvested, the job is not over. Further work is needed to make the cones usable in the brew pot and to preserve those that will be held in storage. This is the task of drying the cones. I have thought about this phase of home growing of hops and have reached the conclusion that two goals are served by drying: uniformity and preservation. A third coincidental goal is involved for some species of hops, namely the European-origin aromatic hops such as Hallertauer, and that is a subtle conversion of the chemical constituents of the yellow lupulin powder inside each cone to enhance the floral components of hop aroma.[5] Indeed, some commercial brewers admit to aging the aromatic hops they use for over a year before committing any to the brewing tanks.

Uniformity is sought through drying so that a given weight of cones will supply similar amounts of bittering from one brew to the next. Consider that a home brewer will likely use an ounce, an ounce and a half, or maybe even two ounces of hops for bittering

---

[5] American Society of Brewing Chemists Journal, Vol. 43 No.3, 127-135, <u>Changes in Hop Oil Content and Hoppiness Potential (Sigma) During Hop Aging,</u> by Robert T. Foster, II and Gail B. Nickerson, 1985.

in a five gallon batch. It would take far fewer green, undried, hop cones to make up that weight than it would of dried hop cones. However, the amount of bittering supplied by the fewer undried cones could be less as less lupulin is available to bitter the brew. This same logic applies to the more volatile aromatic components of the lupulin that are incorporated into the brew by adding aromatic hops in the last few minutes of the boil (to avoid loss through over-boiling). So drying of the cones, hopefully to a uniform degree of dryness, allows the home brewer to control the results, brew after brew.

The goal of preservation is served through drying hop cones, just as that goal is for other foodstuffs we use. Moisture fosters change, through decay, by growth of unwanted molds and wild yeasts, and by chemical change of the item. Drying reduces the likelihood of such unwanted changes in hops just as it does in say dried apricots, or raisins. In the long history of mankind before mechanical refrigeration was discovered and applied, drying was one of the basic ways to preserve food and many other things. It still works. But today, we may combine the virtues of drying with other preserving techniques to achieve even better results. I have in mind freezing. After hop cones have been dried to a satisfactory degree, they may be kept for many months in a freezer. Little change in flavoring performance will occur in hops so stored. However, progress toward that third goal above, the subtle change in floral components by aging will be slowed by freezing. That process is favored by keeping hop cones in an environment devoid of air and light, and cool but not below freezing. Hard conditions to attain for the home grower of hops, to be sure. So, compromises are in order. I have found that long term storage of dried hops, sealed in quality plastic bags and kept in a freezer, is a satisfactory compromise between preservation and aroma improvement by aging. This I can do with supplies and equipment already in the kitchen, and that is an important consideration!

Now let us turn to the how-to-do-it of drying hop cones brought in from the hop yard. The first rule is to handle these delicate objects with care. Be gentle with them. That yellow lupulin powder will sift out of the cone with rough handling, and that powder is the source of all the good flavor that the home brewer seeks to incorporate in the brew. Don't lose the "goodies" of the freshly harvested hop cones by treating them with indifference!

I shall mention three ways to dry the cones. Obviously, other ways exist. One may use sunshine and time, as hops were dried historically. Bugs may land in the exposed hop cones, along with dust and pet hair, and some spoilage may occur before adequate drying has been achieved. I know of some growers who place their hops in an attic to get even higher temperatures than produced by direct sunlight. This works, yes. But for me, I do not wish to risk the hazards involved. I want more control of the process and faster results so the dried hops can be bagged, sealed and stored in the freezer, in the least possible time after harvesting.

If you happen to have a food dehydrator in your kitchen, that appliance may be used successfully to dry hops. Just set the control at around 115°F and check the cones periodically for progress. It may take a full twenty-four hours to dry them, and there will be a distinct hop aroma left in the dehydrator. The slight stickiness on the dehydrator trays will come off with soap and water. The main limitation to use of the food dehydrator is capacity. Unless you are the owner of a rather large one, it will take several sessions to dry even a small harvest of hop cones. Ours is a four-tray dehydrator, a rather small machine. Even a three-pound coffee can full of hop cones exceeds its capacity and requires two loadings. Clearly, for our purposes, greater drying capacity was a must.

The solution I came up with was to construct a hop dryer. I have include data in the appendix for others inclined to do such a project. Luckily for me, there was an old "helmet" style woman's hair dryer in our garage, gathering dust. That became the heat and forced air source for the hop dryer I built. It works wonderfully well. I am able to dry the equivalent volume of six three-pound coffee cans of hop cones in about twelve hours. The dryer has two screen-bottomed drawers, one on top of the other, with the hair dryer mounted on the lid. The top drawer load will dry overnight, then the lower drawer is moved to the top position to finish drying while I package the first load. The second drawer is dry by mid-morning. This is a lot faster than drying with the dehydrator, and certainly quicker than sun drying. I recommend that some sort of volume dryer be included in the plans of anyone contemplating being a homegrower of hops.

At this point, a reader will be entitled to ask the question, how dry is dry enough? Commercial hop growers would answer than most reasonable question in terms of percentage of residual moisture in the cones, as determined by weighing before and after drying. Rehydration to a limited degree would also be mentioned. But, the homegrower of hops needs something more practical, more in keeping with the modest resources available in the average home, to guide the timing of the drying operation.

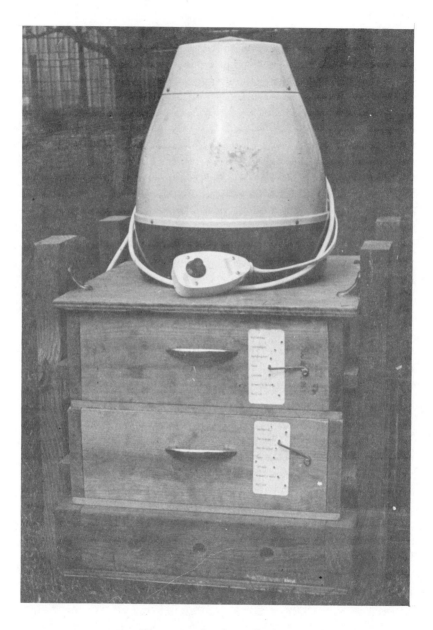

Home-made hop dryer.

I do not claim scientific validity for what I am about to suggest, but I do proclaim its utility as a practical guide to determine "dry enough" to package. This is feel. Dry cones until they are springy to the touch, like popped popcorn. The scales of the cone will have spread open noticeable at this stage. The lupulin powder will fall out easily if the cone is dropped say onto a table top from eighteen inches or so. Carefully scoop up the cones, placing each handful into a good quality plastic sealable bag. Fill the bag and compress the cones firmly, then add more cones to use the full capacity of the bag. When you cannot squeeze more cones into the bag with modest pressure, attach the bag to your sealing machine, compress the bag to force out as much air as possible, then seal and label it. I use a good quality, boilable sealing bag. These non-permeable bags do cost a bit more that the zip-lock style, but are worth the extra cost. The zip-lock style, in my experience, will allow moisture to enter the contained hops. Before too long, the hops take on the look of cooked spinach. I find that not too appealing for use in the brew pot. By contrast, hops stored and frozen in the boilable bags will retain their color and feel for a year or more. It is wise to avoid repeated thawing and refreezing of these stored hops, and do not leave the bags exposed to direct sunlight for any length of time before freezing them. And do take time to label each bag with the type of hop contained and the year of harvest. Failure to do so will assuredly produce real frustration in the future.

There is a good clue to whether or not hops have been dried "enough." This is the presence of dampness inside the bag after the bag is removed from the freezer and allowed to come to room temperature. The feeling of slight dampness to the hand is normal but the presence of beads of moisture or wetness of the cones at the bottom of the bag are signs that the hops were not sufficiently dried before packaging. Get them drier next season is the after-the-fact lesson to be learned from this clue.

CHAPTER 10
Fall/Winter Care

In late fall or early winter, but certainly before hard freezes or snowfall occurs, it is time to say "thank you" to your hop plants for the harvest of hop cones the plants have produced earlier. The way to do this is by providing the care necessary for the hop plants to winter over without damage, and to be ready to renew growth with vigor when spring arrives.

Once the hop leaves have begun to fall from the vines and the vines have started to turn brown, it is time to trim back the old growth to ground level. Long handled lopper shears are handy for this task. Some vines will have grown to over an inch in diameter at the base, so a bit of leverage is helpful to make a clean cut. I have tried two ways of beginning the trim. The first way is simply to climb up on a ladder positioned along side the climbing twine so as to reach the highest part of the old growth, then unwind the vines from the string. When the unwound section becomes too long for easy handling, use pruning shears to cut the section off, and continue in this fashion until all vines on a given twine have been unwound to ground level. Then, use the loppers and cut the vines at ground surface flush with the ground. The second method reverses the process. Begin by using the loppers to cut the vine at

77

ground level.
Cut all vines
growing on a
given climbing
twine, then un-
tie the twine
from the "J"-
bend anchor
wire. This
leaves the bun-
dle of old vines
swinging free.
Working in
about yard-long
sections start-
ing from the
bottom, pick out
and pull free
the twine
around which
the vines have
grown. Contin-
ue, working
upward and
from the ladder
as needed, un-
til the twine is
fully removed
from the vines.
Once freed from

Trimming with loppers.

the twine, the old, full-length vines can be put to good use as will be discussed later.

Whether to use one or the other of these trimming techniques is a matter of individual preference. The top-down, first way risks that due to movement of twine and vine bundle during trimming, some of the hop plant below ground level may be pulled out of the ground. If this occurs, simply trim the vine where it disappears into the soil and hope that enough is left below ground to grow again next season. This top-down method will leave your hop vines cut into sections short enough to stuff into a garbage can for disposal, and may save time in hop yard clean-up after the trimming is done. The loss of full-length vines, however, may not be desired. If the grower has plans to make use of the old vines, the second method may be preferred. The bottom-up, second way of trimming also avoids the risk that portions of the hop plant may be pulled out of the soil during trimming.

Regardless of which trimming technique is used, the grower may wish to cut off and save the lower portion of particularly healthy vines. Choose sections that still show green on the outside, with clusters of short sprouts appearing around the stem every six to eight inches. These sections may be heeled over until early spring, then used to start new hop hills. Simply dig a shallow trench -- three or four inches deep will suffice -- and lay the section to be saved in the trench and cover it over with soil. Be sure to mark the location well, to be able to locate the section in the spring. Pieces of healthy lower vine growth four to six foot long can be held over winter in this fashion, to be cut into foot-long sections the following spring for planting in new hills.

Once the hop hill has been trimmed of old growth, it is ready to receive its over-winter blanket and nourishment. The exposed cut and base of the hop needs protection from the weather, and the roots need feeding to rebuild strength for the new growth the following spring. The mulch that I have used with good results comes from the chicken house. It is a mixture of manure and straw and provides both thermal protection and nutrition for the hop plant during its dormant season. Mulching also helps control weed growth. It can be left in place during the next growing season and allowed gradually to mix into the soil around the hill. I use a layer three to four inches thick over the hill and extending beyond the hill about six to twelve inches. This treatment also gives the hop yard a nice orderly appearance for the winter.

One last task remains to complete winter care. That is tying now-free climbing twines to the poles that support the upper wire of the hop yard, so that the wind doesn't play mischief with the twine during winter storms.

-- Using Old Hop Vines

Earlier mention was made of a use for old hop vines left full-length during the trimming operation. That use is as a core for wreath-making. When first trimmed, hop vines remain rather flexible even though brown on the outside.' Those the size of an ordinary lead pencil are somewhat springy and only slightly more stiff than hemp rope of the same diameter. After becoming thoroughly dry, the hop vine loses its pliability, but still retains its springiness. These characteristics are particularly useful in application as wreath cores. Cut greens, holly, or ornament hooks can be inserted between the coiled hop vines, using the natural tension among the vines to hold inserted materials. Suggestions on wreath core preparation appear in Appendix B.

There is a ready market for wreath cores. They are offered for sale at craft supply stores, and frequently at craft fairs or bazaars. The retail price for cores range from about $5 for smaller ones (about twelve inches across), to nearly $10 for cores two and one-half to three feet in diameter. For the enterprising homegrower of hops who is willing to invest some spare time after the trimming chore is done, what otherwise would be trash for the garbage collector can be transformed into a source of modest income and a useful product for others to complete.

CHAPTER 11
Other Uses for Hops

This chapter presents a real challenge to me, I will admit. This is because I have a very hard time believing that there is any other proper use for hops than as the vital essence of a good beer. Such a stereotyped view is hard to shake. But, I will give it a try.

One use for hops of which I have read is as pillow stuffing for the relief of people bothered by congested sinuses. Now, my sinuses bother me during Oregon's long and very damp winters. Come to think about it, they bother me in summer time too, when field dust, pollen, and who knows what else roils about in the atmosphere. So, maybe I should try stuffing a pillow with hops and sleep with it to learn whether this usage is valid. Until I have made such a trial, however, I cannot vouch for the usefulness or even the wisdom of hops for relief of sinus problems. Readers may investigate this theory at their own risk and expense!

Now for some other uses of hops that more closely parallel the customary application for these fragrant fruits of your labor. I still am fixed on the idea that hops belong in a brew pot, but in the present context, consider that your hops may end up in someone else's brew pot. What I am suggesting to you is this, that hops excess to your needs may be ideal items for trading with other home brewers, for different varieties of hops grown by them, or what have you. Why, you might even consider selling some of your excess harvest. Or, given a cooperative retailer of brewing supplies such as barley malt, swapping your fresh hops for other necessities of the brewing hobby.

We have found another simple and satisfying use for hops. This is in connection with our membership in a brewing club. We have taken bags of hops to meetings for use as doorprizes, or for an occasional club raffle as a minor fund-raiser. So the idea of hops as a gift, either through a brewing club, or to individuals also involved with home brewing, is worth trying once in a while.

As a final thought on this topic of other uses for your home grown hops, remember that your hop plants produce more than just those cones at harvest time. While those cones are forming and maturing, the crown of the hop plant is busily making new rhizomes. There is a market for good, healthy rhizomes. Just get the word out to would-be growers that you will have "some extra rhizomes" in the early spring, and folks will seek you out to start their own hop yards. This is particularly true for the aromatic (and sometimes slow to grow) hops, the rhizomes for which are often not available from commercial sources.

Beyond these thoughts, I shall leave it to the reader's imagination to discover other uses for the delightful, flavorful bounty of the homegrower's efforts at hop cultivation.

APPENDIX A

Home-built Hop Dryer

Why would anyone want to spend to time and effort to build a hop dryer? Beyond time and effort, some minimum supplies are needed as well and that means money to be spent. The answer that I have arrived at, and that I now suggest to others, is one word, necessity. If a homegrower of hops wants to process the results of considerable effort, namely the harvest of hop cones, with relative speed and ease, and in a non-contaminating fashion, then a hop dryer is necessary. This is because the volume of hop cones will be too great to handle satisfactorily without use of specialized equipment. Rather than embarking on a search for a manufacturer of a suitable piece of equipment, one can better invest that time in making the desired device. At least, that is the opinion of this writer.

To that end, I shall give some thoughts on how an efficient hop dryer may be built with a minimum of skill required, very few supplies to be bought, and in a rather short time. I built mine in one evening. But I had a head start in that the drawers were already constructed and required only slight modification. Now it would be useless for me to try to specify exact dimensions; that choice is better left to each builder. I will give the over-all dimensions for the drawers that I used, and for the resulting completed dryer:

Drawer size: 15 1/2 " wide, 20 " deep, 5 1/2 " high
Dryer (using two drawers): 17 " wide, 21 1/2 " deep, 20 " high

I will next cover in some detail certain special design features of the hop dryer that I built, in the belief that such information will be of greater value to others than detailed dimension data.

Drawer detail: Sides made of softwood (fir, hemlock, or pine) 1x6, with a one-half inch strip cut off and saved, leaving the starting width of the sides

at five inches. Bronze window screen cloth was used for the bottom of each drawer, and the one-half inch strip used to secure the screen cloth. Thus, once that one-half inch strip is tacked in place, the completed

drawer is five and one-half inches high. Bronze screen cloth is recommended as it does not rust. Recall, the green hop cones will lose a lot of moisture during drying, so rust-proof screen cloth is a sensible precaution against early replacement. Simple one inch strips were attached by screws to the sides of each drawer, to act as rails and support the drawer in the dryer frame. Be careful to attach these strips at the same place on each drawer, so the drawers may interchange positions in the dryer. I used an ordinary drawer handle on the front of the drawer, to make removal easier. Next to the handle, as the photo reveals, I attached a short piece of galvanized wire bent at one end to form a small eye (for an attaching screw) and at the other end with a sharp, short right angle. I made a slight bend just beyond the eye of maybe 15°, so that once put in place with a screw, the sharp right angle bend was under slight pressure on the wood surface. I then added a paper label on which I had typed the name of each variety of hop growing in our yard. Moving the wire in an arc, I marked positions for boring a "dimple" through the paper label at the places where hop varieties were listed. This arrangement allows the wire to function as a movable pointer, to show which variety of hop is in the drawer at any time.

Dryer frame detail: I used 2x4 lumber for the frame for rigidity. There is no need to make sides as such, because the sides and ends of the drawers fully enclose the dryer. The bottom tray is a 2x4 rectangle

underlaid by a piece of one-half inch plywood. I bored three-quarter inch diameter holes through the end 2x4s, three holes each, to provide for discharge of air. To this bottom tray, I attached 2x4 uprights, two to a side. Between the uprights of each side, I added one inch stock to correspond with the rails attached to each drawer. As a top, I again chose one-half inch plywood. I used a sabresaw to cut out a circle eleven inches in diameter in the center of the top for mounting the helmet style hair dryer. To attach the top to the uprights, I used furniture-style corner braces, one per upright, and left the screw placed into the upright slightly loose. This allowed the top panel to move up-and-down about one-eighth inch, enough to make drawer removal easy, yet permitting the weight of the hair dryer to press the top down firmly on the upper drawer and make a close seal to avoid air loss. Gravity is still effective! A bit of paraffin rubbed on to the touching surfaces of the drawer rails aids movement. The accompanying photos of my hop dryer show how the completed frame looks and will reveal additional clues for construction.

Hair dryer attachment: This presented a bit of a challenge. I knew that some way had to be found to force the warm air to enter the chamber created by the screen-bottomed drawers. The solution proved quite easy. We had an old plastic garbage can that had seen better days. Before putting it in the trash I did some work on it with shears. The result was a strip of heavy but flexible plastic about two and one-half inches high. I had already stapled a piece of fiberglass screen cloth carefully inside the eleven inch hole I had cut out of the top panel. Now I added this strip of plastic, stapling it to the inside edge of that hole. The result was a plastic shroud that extended upwards, into the opening of the helmet hair dryer, making an effective seal to assure the full volume of warm air entered the drawers. The lip of the hair dryer rests on the plywood and thus presses down to make a tight seal with the top, and the plastic shroud prevents the loss of any air. The hair dryer is not permanently attached or otherwise converted

from its original purpose. The diagrams I have prepared will give a better appreciation of this design feature than the photos could offer.

Adjustable pointer

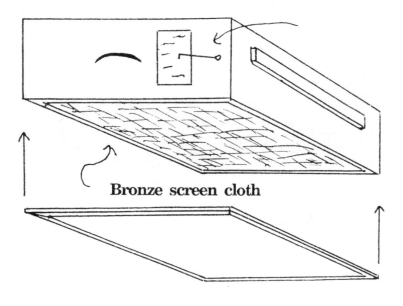

Bronze screen cloth

1/2" strips

<u>Drawer Construction Detail</u>

91

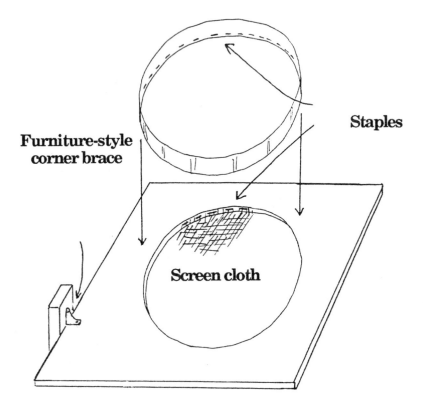

Staples

Furniture-style corner brace

Screen cloth

Lid Construction Detail

Appendix B
Wreath Frame Making Basics

Freshly cut hop vines are almost as pliable as clothesline and remain so for several days after cutting. Once they have dried, however, the vines are quite springy and will retain the form given to them while yet green. These characteristics make hop vines an ideal medium for constructing wreath frames. They are easy to work with, and will hold evergreen boughs or holly that is later worked into the frame as holiday decorations.

The first step to be done is to remove the leaves from the hop vines to be used for wreath frames. Some may choose to leave some of the leaves in place for a more rustic appearance. Cut the vines to a length appropriate for the size of frame to be made, about six and one-half foot long for a wreath two foot in diameter, and about eight foot long for three foot frames. Have on hand a supply of soft annealed iron wire such as used by florists, and a pair of snippers to cut this fine wire.

The simplest type of wreath frame is just a bundle of hop vines, say six or eight strands, held into a circular form by wire wraps at several points around the circumference. It is possible to use the hop vine to make a built-in hanger for the frame by forming, in one vine, a small loop of about two inches and tying the loop securely with wire. Place that section of vine on the outer layer of vines used to form the frame, and position the loop for eventual use as a hanger. Then secure the bundle of vines to hold them in place until dried.

For those who anticipate making wreath frames in quantity, or over the course of several years, it may be worth the time and effort to construct a jig on which to form bundles. This can by done with some dowels and a piece of particle board or plywood. Draw two concentric circles on the board of a size to represent the inside and outside of the wreath frame desired. Then bore holes the diameter of the dowels to be used every three or four inches along these circles. Insert short lengths of dowel, say four or five inches long, with a bit of wood glue and allow to dry. I suggest that the dowels be one-half inch or larger in diameter.

To use this jig, place sections of hop vine into the space bounded by the dowels, varying the beginning and ends of vine lengths, until the size of bundle desired is attained. The dowels will hold the vines in position until the seizing wire is wrapped around the bundle at several points, then the frame may be removed from the jig. An extra dowel may be installed to position the vine hanger correctly, as shown in the diagram.

A slightly more complex technique for frame-making involves the use of hop vine to wrap the bundle and hold it together. Some wire will still be needed to secure the ends of the hop vine used as a wrap, but the over-all appearance of the finished frame will be natural and slightly bulkier than those held together by wire alone.

For those who are more ambitious, or are more creative, wreath frames may also be made by

94

braiding the hop vines. One may make braids of three strands or four, or even more. These can be used directly for slim frames, or braided together with other braids for very full-bodied frames, according to one's wishes and imagination. Done by braiding, wreaths do not require a jig and the diameter of the frame may be varied from one to another easily. It is a good idea to use something to clamp the ends of the vines to be braided, to hold them so that the braid can be made. A bench vise may do the trick, or a large carpenter's wood clamp. As the vines will not all be of uniform diameter, some crushing of the very ends will necessarily occur to hold them in place firmly enough to be able to work with them. Add enough extra length to allow this portion to be cut off after the braid is completed. The truly ambitious will form the circle for the wreath frame by braiding the ends together, like a fisherman splicing a rope. Others may decide to use the soft iron wire to complete the task.

For the sake of understandability, the braiding illustrations to follow will use "color" names to identify the various strands of hop vine to be manipulated:

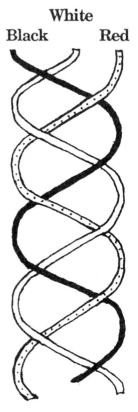

White
Black   **Red**

Black over White

Red over Black

White over Red

Black over White

Red over Black

White over Red

Black over White

Red over Black

White over Red

Black over White

and so on . . .

Braid of Three Strands

[W]hite   [R]ed
[B]lack   [Y]ellow

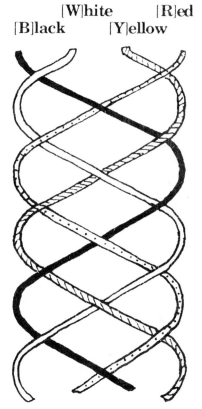

B under W   Y under R

B over R

W under R   B under Y

W over Y

R under Y   W under B

R over B

Y under B   R under W

Y over W

B under W   Y under R

B over R

and so on . . .

Braid of Four Strands

Loop dowel

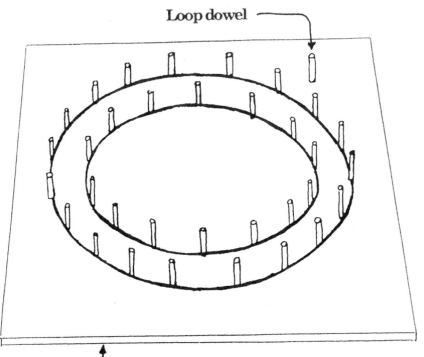

3/4- to 1-inch stock

Wreath Frame Jig

INDEX

V

verticillium wilt   42-43
vigorous growth   55
vines, training   56-58
viruses   41-43

W

water   5, 26, 41-44, 52
weeds and weeding   40, 51
Willamette Valley   19, 61
wilt, verticillium   41-43
wind   25, 48
winter care   78-80
wire (attachment)   7-10
    (guy)   9
    ("J" bend)   58, 78
    (top)   8
withering   39
wreaths   80-81, 93-98

X-Y

yellow, leaves   42, 46-47
young growth   32, 42
youngsters   27

Z

zinc   47
zineb   42